FRUITS YOU LOVE TO EAT
APPLES

AMY CULLIFORD

A Crabtree Roots Book

Crabtree Publishing
crabtreebooks.com

School-to-Home Support for Caregivers and Teachers

This book helps children grow by letting them practice reading. Here are a few guiding questions to help the reader with building his or her comprehension skills. Possible answers appear here in red.

Before Reading:
- What do I think this book is about?
    - *I think this book is about how good apples taste.*
    - *I think this book will tell me how important it is to eat apples every day.*
- What do I want to learn about this topic?
    - *I want to learn more about how apples grow.*
    - *I want to learn why some apples are red and others are green or yellow.*

During Reading:
- I wonder why...
    - *I wonder why apples come from little brown seeds.*
    - *I wonder why apples taste so good.*
- What have I learned so far?
    - *I have learned that apples grow on trees.*
    - *I have learned that apples are fruits.*

After Reading:
- What details did I learn about this topic?
    - *I have learned that all apple trees have green leaves.*
    - *I have learned that brown apple seeds grow into apple trees.*
- Read the book again and look for the vocabulary words.
    - *I see the word **fruits** on page 3 and the word **leaves** on page 8. The other vocabulary words are found on page 14.*

Apples are **fruits**.

Apples come from little brown **seeds**.

The seeds are put in the ground and grow into **trees**.

All apple trees have green **leaves**.

Apples can be green, yellow, or red.

Crunch! I like
to eat apples!

Word List
Sight Words

all	crunch	I	red
and	eat	in	the
are	from	into	to
be	green	like	yellow
brown	ground	little	
can	grow	or	
come	have	put	

Words to Know

apples

fruits

leaves

seeds

trees

39 Words

Apples are **fruits**.

Apples come from little brown **seeds**.

The seeds are put in the ground and grow into **trees**.

All apple trees have green **leaves**.

Apples can be green, yellow, or red.

Crunch! I like to eat apples!

FRUITS YOU LOVE TO EAT
APPLES

Written by: Amy Culliford
Designed by: Rhea Wallace
Series Development: James Earley
Proofreader: Melissa Boyce
Educational Consultant: Marie Lemke M.Ed.

Photographs:
Shutterstock:Natalay Studio: cover; CGIHeart: p. 1; Lightfield Studios: p.3; Ermak Oksana: p. 5; Paul Hardwick Images: p. 9; grey_and: p. 11; Monkey Business Images: p. 13

Crabtree Publishing

crabtreebooks.com   800-387-7650
Copyright © 2024 Crabtree Publishing
All rights reserved. No part of this publication may be reproduced, stored in a retrieval system or be transmitted in any form or by any means, electronic, mechanical, photocopying, recording, or otherwise, without the prior written permission of Crabtree Publishing. In Canada: We acknowledge the financial support of the Government of Canada through the Canada Book Fund for our publishing activities.

Printed in the U.S.A./072023/CG20230214

Published in Canada
Crabtree Publishing
616 Welland Ave.
St. Catharines, Ontario
L2M 5V6

Published in the United States
Crabtree Publishing
347 Fifth Ave
Suite 1402-145
New York, NY 10016

Library and Archives Canada Cataloguing in Publication
Available at Library and Archives Canada

Library of Congress Cataloging-in-Publication Data
Available at the Library of Congress

Hardcover: 978-1-0398-0973-4
Paperback: 978-1-0398-1026-6
Ebook (pdf): 978-1-0398-1132-4
Epub: 978-1-0398-1079-2